For Delilah, Teddy and Fliss

ISBN: 978-1-7384235-8-3

First published in 2024 by Institute of Physics, 37 Caledonian Road, London, N1 9BU.

This book was made possible thanks to the generous support of the UK Space Agency through a *Space for All* grant.

About the Author

Toby Shannon-Smith works at the Institute of Physics and lives in West London with his husband and his two cats, Momo and Appa. When he's not writing books or attempting to stop his cats chewing his laptop charger, you can probably find him painting, crafting or cooking. He has a degree in physics and an MSc in Science Communication. He previously wrote ***Mimi's Rainbow Adventure*** and this is the next one in the series!

About the Illustrator

Pauline Gregory is a children's illustrator living on the Cornish coast. Following her passion, Pauline graduated from Arts University Plymouth with a First Class Honours BA in illustration. She takes inspiration from the funny and endearing things that animals and people do everyday and the superb imagination that children have. Pauline works with traditional materials to create quirky illustrations with humour and charm.

Graphic Design

By Jack G Sheppard

Science Superstar Fact Files

Edited by Imogen Thompson

With thanks to

The Public Engagement team at the Institute of Physics:
Steph Bevan, Sammi Bowers, Brendan Owens and Louise Swan.
And all of the amazing physicists, scientists and engineers that bring Mimi's Space Adventure to life!

Notes for Parents, Carers and Teachers

Hello and welcome to *Mimi's Space Adventure!*

We hope that you and your child, family or class enjoy reading this story together and being part of Mimi's latest adventure. As you read the story, you'll meet lots of fun characters who are doing amazing jobs relating to space – make sure you keep reading after the end of the story to meet the real-life scientists, engineers, apprentices and technicians who are making a space adventure a reality!

At the Institute of Physics, we want to make sure that no child misses out on the opportunities that doing science, and especially physics, can bring. By doing things together, like reading this book as a family, you're supporting your child's curiosity and keeping their options open about the future.

To see the stories behind real-life science role models, find free science activities to do together and support your child's aspirations, visit the Institute of Physics website at *www.iop.org/SpaceAdventure*

Meet our Heroes!

Mimi is a little girl with a lot of big questions! She is always asking Why? How? Who? Where? When? Mimi wants to know all about how the world around her works.

Granny is one of Mimi's favourite people. She loves to invent and create things in her garage and has always got a new project on the go.

Colin is Granny's dog and loves to go with her and Mimi on different adventures.

DAVE is a robot Granny invented that always seems to know the things people need and the best way to find them.

A little while ago...

Mimi sharpened a pencil and got a fresh piece of paper. She looked up at the poster of the Solar System she had pinned to her bedroom wall and an idea came into her head.

Thinking carefully, she poked her tongue out of the corner of her mouth and, in her best handwriting, started to write...

Once she had finished, she put her paper in an envelope and added a stamp. She put it in the post box and then forgot all about it and got on with enjoying the end of term at school...

07:00
Chapter One

Chapter 1

It was the summer holidays, again.

Mimi's Mum and Dad were away and her brother, Dylan, was working as a chef in the big city. This meant that she was spending her holiday at her Granny's house with Colin the Dog and DAVE the Robot.

She spent her days happily busy; helping Granny build inventions in her garage, going on trips on her solar-powered motorbike and taking Colin for long walks in the fields and woods near Granny's house.

As ever, she was full of questions about the world around her. How did a cake go from runny cake mix to fluffy sponge, how do DAVE's circuits and machinery work and why does Colin love chasing his own tail so much?

It was a nice summer holiday but soon it was going to become a great adventure once again!

One morning, at breakfast, Granny was stirring a mug of tea with a screwdriver whilst her attention was on the newspaper crossword and Mimi was tucking into a raspberry yoghurt.

"Woof?" said Colin, nervously. Mimi looked over and, with a loud clatter, she dropped her spoon on the table. As she looked out of the window she saw a large figure walking slowly up the garden path to the front door...

The doorbell rang and Mimi cautiously opened it a crack... She peeped out and standing on the other side was a hulking figure. It was wearing a white suit with it's face hidden behind a gold-coloured visor and a fishbowl-shaped helmet that covered their entire head.

Huge gloves and boots were on their hands and feet and they were holding an official-looking envelope.

"Umm... Hello?" she squeaked, and then, thinking of the pictures she had seen in some of her books: "Are you... an astronaut?!"

The figure didn't answer but replied with a question of their own:

"Are you Mimi?" it said, its voice distorted by the helmet, making it sound like someone talking into a metal bucket. Mimi nodded, nervously.

The figure put a hand to their helmet and the gold visor rolled back to reveal a friendly, smiling face inside.

“Here ya go!” they said and handed Mimi the official-looking envelope. “See you soon!” they said as they started to clomp back down the path in their heavy boots and gave her a little wave. They got into a big, impressive-looking car which drove off with a squeal of it’s tyres and a cloud of dust.

Chapter
Two

07:10

Chapter 2

Mimi was baffled. She closed the front door and looked down at the envelope in her hands. It was big and made from thick paper and had her name on the front in large, official letters.

"Was that the postie?" asked Granny, not looking up from her newspaper.

"Umm... Sort of!" replied Mimi, turning the envelope over in her hands. She sat back down in her chair and looked at the envelope curiously.

"Ooh, anything interesting?" asked Granny, as she triumphantly wrote down an answer to one of the crossword clues.

"A letter, to me!" said Mimi. She wiped jam and crumbs from a butterknife and slit open the envelope carefully.

There was a loud POP noise that made Colin run underneath the table. Granny's head jerked up from her crossword to see many little coloured squares of paper and glitter raining down around the kitchen. Intrigued, she fished a piece of confetti out of her tea and went to join Mimi to have a look at the letter.

CONGRATULATIONS

Dear Mimi,

We **LOVED** your essay, 'Everything I've Ever Wanted to Know About Space'.

We'd like to invite you to an all-expense-paid, carbon-neutral, once-in-a-lifetime visit to Mission Control! You'll meet astronauts and space scientists from all around the world and get a unique look at what it takes to go to space!

Later on this week, we'll collect you, and up to two adults of your choice, to go on your big space adventure! We look forward to welcoming you and kids like you from all over the world! Pack a bag and get ready for an incredible experience!

"OH MY GOSH!" yelled Mimi as she and her Granny did a happy dance around the kitchen until they got dizzy and out of breath.

Flopping back down in their kitchen chairs with the confetti swirling around them, Granny took another look at the letter.

"Oh dear..." she said, pointing at a line that was hidden below a fold in the paper.

PS – sorry, no pets allowed.

Colin poked his head out from underneath the table and huffed, moodily.

"Woof," he said, sulking.

Mimi was feeling sad that Colin couldn't join them on this amazing adventure! It wouldn't be the same without him and she knew how much Granny hated leaving him behind... But then she had an idea! Whispering to DAVE, he told her where to look to find what she needed.

Mimi led Colin to the garage and she rummaged through the piles of boxes, junk and inventions until she found a box that was labelled **AMATEUR DRAMATICS SOCIETY**.

Picking through the box, she looked until she found the costumes from their most recent production.

Giggling, she walked back into the kitchen where Granny was still picking bits of confetti out of her hair and her breakfast.

"Granny, presenting... **'UNCLE' COLIN!!!"** She opened the kitchen door with a flourish and revealed Colin dressed in a big coat, hat and sunglasses and a false moustache stuck to his furry face.

"Woof?" he said, uncertainly.

Granny and Mimi immediately fell about laughing. "Mimi, you know what?" said Granny, wiping the tears of laughter from her eyes as she got her breath back. "This might just work!"

CHAPTER THREE

Chapter 3

Finally, the big day arrived. Their bags were packed and waiting by the front door. They heard a car pull up outside Granny's house and a horn beeped twice and they opened the door to see a suited driver in a peaked hat standing on the doorstep and a big black car shining in the morning sunlight.

The driver did a smart salute and said good morning as they collected the suitcases and walked down to the car, opening the doors for Mimi, Granny and 'Uncle' Colin and putting the bags inside.

DAVE waved them off from the doorstep – he couldn't fit in the suitcase and, besides, someone needed to water Granny's tomatoes in the greenhouse! He was also secretly looking forward to watching game shows on daytime television and doing a jigsaw.

Settling back on the plush seats, Mimi and Granny exchanged an excited look as the car's electric motor hummed into life and they drove off down Granny's street and onto the motorway... Colin was going to spoil his disguise by poking his head out of the car window with his tongue hanging out until a stern look from Mimi made him sit down quietly and behave.

"Woof." He said, quietly.

"Bless you!" said the driver, offering him a tissue.

After a short drive, they got to the airport where they were whisked away though secret VIP corridors to a waiting plane. They had fun in first class, where Mimi got to ask as many questions about the plane as she liked as the other passengers filed on board.

Once everyone had found their seats and packed away their bags, they flew up, up, up into the sky...

After they landed, they were collected by another car and eventually ended up at a posh hotel. The hotel's manager met them at the front door and welcomed them in, bringing them to a room on the top floor.

There was a big bowl of fruit on the table and helium balloons that spelled out "Welcome Mimi!" next to a squishy sofa, a giant television and doors leading to three bedrooms where someone had made their bath-towels into the shapes of animals and birds which Colin woofed at, dubiously.

Mimi and Granny explored the hotel room, jumping on the beds and looking in all the drawers and cupboards. After a hot bath with lots of bubbles, Mimi finally got into bed and fell asleep before her head hit the pillow – what an exciting day!

Chapter Four

Chapter 4

Mimi and Granny woke to a knock on their hotel room door. The hotel manager was there with a gift-wrapped parcel that he handed to Mimi.

Opening it excitedly, she unwrapped a brand-new jumpsuit covered in pockets, badges and patches, with her name embroidered on the front. She was going to look like a real astronaut!

She carefully stowed her lucky compass, her magnifying glass and plenty of pens and pencils into the jumpsuit pockets so that she would be ready for anything and made sure to pick up her Big Book of Questions to take along with her.

After breakfast, Mimi and Granny were picked up by another shiny car and driven out into the middle of nowhere... They went past lots of official looking security checkpoints and, eventually, stopped outside an impressive building with hundreds of flags from different countries fluttering in the breeze outside. Mission Control! At last!

She was welcomed through the door and joined a group of other young people wearing jumpsuits just like hers who were all chatting excitedly. The grown-ups were shown to a comfy sofa where they looked on proudly.

Feeling a little bit shy, Mimi decided to be brave and introduce herself to the other children. The others had come from all over the world and all of them loved space as much as she did. Of course, she had a million questions for all of them but they stopped talking when a person came up to a microphone and started to speak to them.

Mimi realised with a start, it was the person in the astronaut suit that she'd met on Granny's doorstep! She was looking very different in a smart suit with a space-themed t-shirt on underneath and some seriously cool trainers.

Luna

Let's Meet… Luna

My job is… Head of Mission Control

Three words that describe me… Fun-loving, energetic and busy!

A fun fact about me… I've actually been to space! I spent three months working on the International Space Station

The best thing about what I do is… I get to work with brilliant people from all around the world to make missions to space happen!

"Welcome everyone!" said the astronaut. "My name's Luna and I am your host here at Mission Control!"

"Hi Luna!" the children chorused.

"I am so excited that you're all here today – you've come from all over the world to share this experience and it's so great to see you all. I hope that you had a pleasant journey getting here and that you're ready for some busy days of science, technology, excitement and discovery!"

The children cheered and Mimi did a little happy dance – she couldn't wait!

"Each of you wrote an incredible essay that asked fantastic questions about space and this is your chance to find out some of the answers! We'll be exploring every part of Mission Control and meeting some of the people that make our space missions a reality."

“And, best of all, your time here will finish with the launch of our latest mission into space! Our astronauts are ready and waiting for blast off and you’ve all got VIP seats to wave them off!” she paused while the children cheered again.

“Well, that’s about enough talking from me! Let’s go! Grown-ups, have a lovely day at the Mission Control Spa!”

“Bye Mimi! Have fun!” called Granny, waving.

“Woof!” said ‘Uncle’ Colin.

“Bless you!” said one of the other grown-ups, offering Colin a tissue from her handbag.

The children followed Luna out of the foyer and began their tour. Mimi’s head was already buzzing with excitement...

CHAPTER

FIVE

Chapter 5

Luna led the group to a cloakroom where they put on white coats and slid covers over their shoes and hairnets over their heads.

They went through a sturdy metal door and squeezed into a tiny room where they were blasted with jets of air that made everyone giggle in surprise. Luna explained that this was to make sure that everyone was completely clean and no outside dust or dirt was going to contaminate the experiments going on inside.

Another door opened and they stepped out into a giant domed room. The air was heavy with moisture and it smelled of green plants and damp earth. Mimi could see scientists hard at work around the room.

They were measuring little green shoots that were growing out of the red-coloured soil on the floor, looking at leaves under a microscope and carefully tending to the racks of seedlings growing with bright lights shining above them.

A smiling man waved to them and walked over. Along with his lab coat, he wore wellies and a toolbelt full of gardening equipment. He wiped his forehead with his hand, leaving a streak of mud on his face and took off his gardening gloves to shake everyone's hand.

Baxter

Let's Meet... Baxter

My job is... Head of Botany Research

Three words that describe me... Messy, friendly and funny

A fun fact about me... Other than gardening, my favourite thing to do is turning the fruit and vegetables that I grow at home into meals that I share with my friends and family

The best thing about what I do is... I've always been curious about how plants grow in some of the most extreme environments on Earth and now I'm trying to grow plants in space! It doesn't get more extreme than that!

"Welcome to my space garden! I'm Baxter and I'm in charge of botany[1] here at Mission Control! Here we try to work out how to make plants grow in space!" he said, gesturing around them. "This experiment is trying to grow plants that will live on Mars."

"Oh!" said Mimi, looking at the soil and remembering the rusty red photos she had seen of Mars. "Is that why the earth is red in here?"

"Spot on!" said Baxter. "We have done our best to recreate it from the information sent to us by the robots that have visited Mars and make it here on Earth! Mars has lots of dusty red soil and the right gases that plants need to grow... So, we've added water and fertiliser and now we're seeing what happens!"

The children all looked around, impressed.

---

[1] *Say "bot-ann-ee" – the study of plants.*

"One day, we might have people living on Mars!" continued Baxter. "If so, we wouldn't want to have to send a rocket every time they needed feeding – food and water are very heavy, after all. We want to help future space explorers to grow their own food and make the most of what they find there."

They talked about the incredible forces that it takes to persuade a rocket to get off the ground and into orbit.

Mimi found out why rocket scientists need to be really careful when they choose what goes into space because every little bit of added weight needs a huge amount of fuel to be added as well. This is why you need to learn how to make your own food whilst you're in space!

Baxter let the group explore the botany lab and chat to the scientists. Mimi stayed behind as, of course, she had lots of questions to ask him!

"Mr Baxter?" she asked.

"Ha! Just Baxter!" he replied. "How can I help?"

"I have always wondered; how can people live on another planet? What would they need to stay alive? Apart from plants, that is." She asked, curiously.

"Excellent question! Well, human beings are like my plants here and every other living thing. To start with, we need food, water and air. We also need things like power so that we can keep at a comfortable temperature."

Mimi nodded – she wouldn't want to be without any of those important things!

"Power is easy enough –"

"Solar panels!" Mimi interrupted – thinking of her Granny's solar-powered motorbike.

"Ha!" Laughed Baxter. "Indeed! And we can grow food indoors like my space garden here if we have water and air... But how do we get our hands on them? There's not much of either on Mars!"

"Hmm..." said Mimi, frowning.

"On Mars, there are loads of signs there was once water. There are things that look like dried riverbeds and the remains of ancient oceans but they're all completely dry. Frozen water, ice, seems to exist at the North and South poles of Mars and scientists have very recently found places where it might be buried..."

"In fact, there's so much ice on Mars, people now think that, if it melted, it could cover the whole planet in a layer of water a couple of metres deep!"

"Like a planet-sized swimming pool!" Said Mimi.

"Yes! But I wouldn't fancy swimming there, it is very chilly on Mars and not much air to breathe either! But we can use ice to help with that, too! We could melt the ice to make liquid water to drink and to grow plants but we can also use it to create air!"

"Wait, you can make air from water?!" asked Mimi.

"Well, the most important ingredient of air for people, anyway! The recipe for water is two parts hydrogen to one part oxygen and, if we use electricity from our solar panels, we can break them apart and collect the oxygen. Even more handy is that we can use the hydrogen and some of the oxygen to make rocket fuel to get our space explorers home again!"

Mimi counted on her fingers – food, check; water, check; air, check; power, check; heat, check... all the ingredients were there!

She thanked Baxter and finished looking around the lab. She watched earthworms burrowing their way through soil and helped scientists to count the leaves on sprouting plants.

Brushing mud off her hands and peeling off her shoe covers, she joined the rest of the group as they said goodbye to Baxter and his team of space gardeners and carried on with their tour.

Chapter Six

Chapter 6

Luna opened another door and the group walked in. It was very dark. Mimi edged forwards carefully with her hands in front of her, not wanting to bump into a wall! She could hear the other children shifting nervously and, with a click, Luna closed the door behind her.

It was completely black in every direction... Mimi felt as though she was floating in space and it was only the reassuring carpet under her feet that told her otherwise!

Mimi was suddenly aware of a point of light above her. And another. And another! Little specks of white light were appearing and settling into the constellations[2] that she recognised from the pictures in her books... The children started calling out the stars they recognised:

"The North Star!"

---

[2] *Say "conn-stell-ay-shuns", the name given to the groups of stars that we can see in the night sky*

"That one's Betelgeuse[3]!"

And then, the Sun appeared, glowing brightly and seeming to hover in the middle of the room. By the miniature sun's light, they could see a smiling person underneath it, who waved at them.

Feeling a lot more confident now they could see where they were going, the group joined them.

---

[3] *Say "bet-ell-jezz" – it's one of the brightest stars in the sky and a red supergiant star in the constellation of Orion*

Melody

Let's Meet... Melody

My job is... Mission Planner

Three words that describe me... Determined, creative and organised

A fun fact about me... I am the captain of my wheelchair basketball team! I love playing sports with my friends and I especially enjoy the parties we have when we win!

The best thing about what I do is... I use loads of different skills to make sure that missions into space go to plan and there's always a new challenge!

“Hello everyone! I’m Melody and I’m a Mission Planner here at Mission Control!” she said.

“Hi Melody!” said the children.

“Welcome to our mission simulator! Or, as I like to call it… The Space Dome!” she said, dramatically.

She pressed a button on the remote control she was holding and they seemed to whizz through space until the dome was filled with an amazingly detailed projection of the Earth. Mimi ducked as a miniature version of the International Space Station whooshed past her head.

“In this room, we use the very latest 3D projection technology to visualise our space missions. We can use the data from computer simulations, telescopes and space probes to create some amazing pictures of space and see how our mission plans will work out in real life!”

The projection shifted again and an image of the Moon filled the dome. Mimi stared open-mouthed at the cratered surface right above her head.

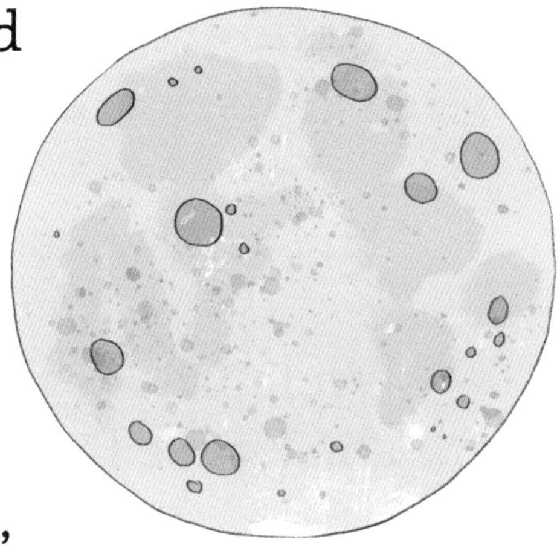

"If we are going to send a mission to the Moon or to Mars or if we want to put a satellite in orbit around the Earth, we need to come up with a plan to get it there! After all, space is very big and you don't want to get lost or end up in the wrong place! It's my job to plan the missions so that we can work out the best way to send our spacecraft to where they need to go."

Wow! Said Mimi to herself as she thought about all the things she had already seen and how hard everyone is working at Mission Control... This sounded like a very responsible sort of job!

Mimi thought for a moment more ...

“But why do you need to plan a mission like this?” She asked. “Isn’t all of space just... Sort of... Up there?” she pointed at the ceiling.

“Ha-ha, yes!” laughed Melody. “Space is all just there but it’s actually all around us! We need to think in 3D! And we also need to remember that everything is moving, all the time! The Moon orbits the Earth and Earth and Mars are orbiting the Sun... All at different speeds!”

Aha. Thought Mimi. This sounded like it was going to be complicated!

“One of the biggest things we need to think about is fuel.” Said Melody. “We can’t just blast off into space, decide where to go and hope for the best! We carefully plan out our journeys to be as efficient as possible...”

Melody clicked her remote again and Mimi could see the Earth, the Moon and Mars all floating together above her.

"Let's imagine we want to go to Mars – when should we go?" she set the planets orbiting around the sun with another click. The planets danced around above Mimi's head, spinning and orbiting at different speeds. Sometimes Mars and Earth were far apart from each other and sometimes they moved closer together...

"There!" one of the other children said, pointing, when the planets were at their closest point to each other.

"Very good!" said Melody. "That is exactly what we need to work out – when are we closest to our destination? We can use lots of data, images from telescopes and maths to work out exactly when that is! And then what do you think we need to do?"

Mimi thought about when she would go on road trips with Granny on her electric motorbike – once they knew where they were going, they needed to be sure they could get there! They needed to charge up the battery!

"Fuel!" she said. "We need to have enough fuel to get there!"

"Yes, exactly!" said Melody. "We need to make sure that the spacecraft has enough fuel to fly to where it's supposed to go! But we can't just take loads and loads – why do you think that is?"

Mimi thought again and, remembered what she had talked about with Baxter: things like fuel and water are heavy!

"Do we need it to be as light as possible?" asked Mimi.

"Very good!" said Melody. "Yes, we need our rocket to be as light as possible when it takes off and we don't want to be weighing it down with extra fuel if we don't need to."

Mimi frowned for a moment.

"Wait... What if things don't go as you planned? Would you run out of fuel?"

"Well, yes... But we think about that, too!" said Melody. "We know that things don't go completely to plan – you might need to make tiny little changes on the trip to stay going in the right direction. We make sure that there is enough fuel to do these changes to the speed and direction of the spacecraft!"

Mimi nodded, this sounded very reassuring!

"After all, on the Apollo 11 mission when people first went to the Moon, they thought they had the perfect place to land... Except it turned out that it was full of rocks! So, the astronauts took over the control of the landing and found a nice flat place to land – and they only had 30 seconds worth of fuel left!"

"Wow!" said Mimi.

"I know!" said Melody. "Space is so interesting – we can use all of our data and simulations like this, here, on the Earth but, there's always something unexpected! We just need to use our best thinking to plan ahead and make sure that we're ready for anything!"

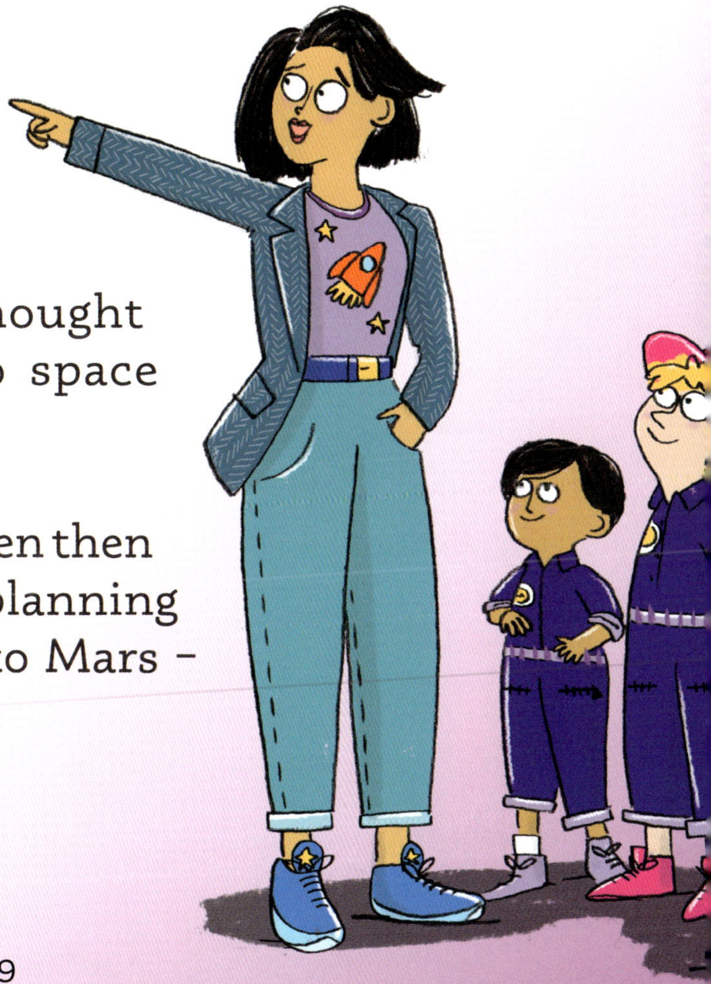

What an adventure! Thought Mimi. Planning a trip into space sounded really exciting!

Mimi and the other children then had a go using the mission planning computer to create a flight to Mars – it was not as easy as it looked!

Chapter Seven

Chapter 7

Luna and the group paused at another doorway and they all put on safety glasses and hard hats before entering another enormous room...

In every direction that Mimi looked, she could see pipes, tanks and machines of all different kinds. Scientists and technicians were taking samples of liquids and opening valves that hissed loudly and sent clouds of icy vapour rolling across the floor like at a rock concert.

Out of a cloud of mist stepped a figure that greeted them.

Rosie

Let's Meet... Rosie

My job is... Rocket Fuel Development Technician

Three words that describe me... Imaginative, careful and cheerful

A fun fact about me... My girlfriend and I have a pet lizard called Frankie!

The best thing about what I do is... I have always loved space so getting to work here is a dream come true!

"Hi everyone, I'm Rosie and welcome to the Rocket Fuel Lab! Here, we test out new ideas for powering our spacecraft to make them faster, more powerful and, especially, more environmentally friendly!"

"Hi Rosie!" they said as she gave them a cheery wave.

"Let me tell you a quick story." Rosie continued. "For the Apollo 11 mission, when the first people walked on the Moon, they used the biggest rocket ever made: the Saturn V[4]. To get it off the ground, it had to burn 18,000kg of fuel... every second! And it took them about two and a half minutes to get up to 68km in the air and off into space."

Mimi gasped as she tried to do the maths – those were some very big numbers!

---

[4] *Pronounced Saturn Five – V is the Roman numeral for the number 5 and it is named after the planet Saturn*

"And, once those rockets were empty, they just separated from the spacecraft and fell back to Earth and – splash! – fell down into the sea. Growing up, I didn't like the idea of how much fuel this used up and then dropping giant bits of litter into the oceans! This inspired me to work on rocket engines that are more environmentally friendly, without using fossil fuels, and that can be re-used so we don't have to waste loads of materials on rockets used only once and then thrown away."

Mimi was amazed. She had seen pictures of these giant rockets in her books, the size of buildings. She never imagined what had happened to them when they left the ground!

Rosie the rocket fuel creator took the group on a tour of her lab. Weaving their way around the tanks and the maze of tubes and pipes, Mimi started to understand how complex getting a rocket off the ground was! As the group wandered around, she had the chance to ask some of the big questions she had in her head.

"So, how are you going to make space launches better for the environment?" she asked Rosie.

"Even modern rockets that might take us back to the Moon or to Mars in a couple of years use a lot of fossil fuels, made from oil. Every time we burn up fuel to do a space launch, this produces about the same amount of gases that a petrol-powered car would do if it drove all the way around the Earth... 70 times over!"

"Wow!" said Mimi. That would be quite the road trip!

"And we also need to make sure those rockets can be re-used as much as possible before they need replacing. After all, you don't throw away your bike after one journey, so why should space travel be any different! And if you're re-using bits, that's much better for the environment and we can also make sure they're not just abandoned in space." Rosie and Mimi laughed together.

Mimi thought back to her chat with Baxter in the space garden...

"Can we use hydrogen to power rockets? Is that any better than fossil fuels?" she asked.

It was Rosie's turn to be impressed.

"Wow! You have been doing your homework!" she said.

"Yes, that's one of the things we're doing here in the lab – using power from the sun to split apart water into hydrogen and oxygen and working out how to use them as rocket fuel. And the amazing thing is, when you burn them together, it just turns back into water!"

Water-powered rockets! Thought Mimi.

"Unfortunately, today's rockets are still running on fossil fuels which is not great for the planet... But the future could look very different! We could be making rocket fuel to get us to Mars out of the dust we find on the Moon! Or powering rockets with solar panels!"

"Coooooool..." said Mimi. This was sounding very futuristic indeed! She couldn't wait to see what happens next!

"Are you looking forward to Launch Day?" asked Mimi.

"Yes! We are all so excited!" squeaked Rosie. "It's going to be amazing to see all our hard work come together as we send our crew into space! I work with people who have come from all over the world who do loads of different jobs and it will be wonderful when it all pays off!"

Mimi couldn't agree more! Buzzing with excitement about the future of space travel, Mimi's group finished exploring the Rocket Fuel Lab and they went to the next part of their adventure...

CHAPTER EIGHT

Chapter 8

Chatting away happily, Luna led the group down long corridors lined with doors. Peeping through the glass, they could see people hard at work on computers, looking carefully at complicated pieces of machinery and having discussions over giant cups of coffee. At each room they passed, people noticed them and they stopped to wave cheerfully.

Getting to the end of the corridor, Luna opened a door and showed them into a room that seemed to be covered in large screens, each showing different kinds of information. Some showed pictures of satellites, flying in graceful curves around the Earth. There were many maps of the stars and lots more had ever-changing sequences of numbers and data that made Mimi feel dizzy trying to keep up!

Luna led the group on a tour around the room and introduced them all to a number of the people working there. Mimi had some big questions so Luna brought her to one of the people working at a computer. Luna tapped them on the shoulder and they turned around.

Tai

Let's Meet... Tai

My job is... Space Debris Tracking Centre Manager

Three words that describe me... Observant, caring and thoughtful

A fun fact about me... I have Attention Deficit Hyperactivity Disorder, or ADHD, which can be challenging sometimes but, now I've found ways to play to my strengths, I feel like I can do anything!

The best thing about what I do is... The issue of space junk is so important for the modern world so it's great to feel like I am making a difference!

"Oh! Hello! Gosh, sorry! It's very easy to lose track of time when you're looking into space! I'm Tai and I'm in charge of all this!!" they said as they waved their hands around the room.

"What we do in here is track the different things that are orbiting the Earth – from satellites to comets, bits of rock and loads of other things that are in the sky above us. It's my job to make sure that our satellites aren't in danger and, if they are, to see if we can do anything about it." said Tai.

"You said that you need to keep satellites safe from danger... Is space dangerous?!" Mimi asked, concerned. Tai saw her frown and smiled.

"Short answer, yes, a bit! One of the big problems that we're tackling here is space junk."

Mimi giggled, imagining old washing machines, tatty newspapers and empty tins of beans in orbit of the Earth.

"It's actually no laughing matter!" said Tai, laughing as they pretended to be serious and failed. "We've been sending stuff into space for the best part of the last 70 years and it is getting full of bits and pieces – rocket parts, old satellites, stuff that has broken off or been lost..."

"Loads of it is orbiting the Earth! At the moment we can track the stuff that's bigger than 10cm or so – about the size of your hand. And there's about 35,000 pieces that we know of! And then there's the smaller bits between 1 and 10cm – there's about a million of those. And then the really small pieces that are under a centimetre... Well, there might be over 130 million!"

Mimi gasped – she'd had no idea!

"Sure, the bits are small but there are lots of them... And they move incredibly quickly! And if they were to hit something important up there... well... that would be very bad news indeed!"

Mimi was quiet. This sounded like a big problem for anyone going into space! Tai patted her on the shoulder.

"Don't worry – this is why I'm here! And the good news is that we're getting better all the time at tracking these bits of space junk and working out how to capture them and return them to Earth. Like doing a litter-pick in space!"

Mimi smiled, relieved.

"How can we get rid of this space junk?" asked Mimi.

"Well, to start with, we have to make sure that we aren't making the problem worse. Now people are more aware of space junk, we can make sure that future space missions don't leave things behind."

"By re-using the rockets?" said Mimi, remembering what she had talked about with Rosie.

“Absolutely!” agreed Tai. “If you are re-using the rockets and other bits of equipment, you aren’t leaving them up there to clutter up the space around the Earth!”

“But what do we do about the stuff that’s already up there?” said Mimi.

“As for the things that are already there, they will eventually fall back to Earth and burn up like a shooting star before they hit the ground but this will take a long time,” replied Tai. Mimi looked unconvinced. Surely there was something that could be done? Seeing the look on her face, Tai continued:

“But that is, of course, way too slow. There’s loads of exciting work going on to solve the problem faster!

If satellites have finished their job in space, and still have some fuel left, we can send them instructions to fly back into the atmosphere and burn up without risking any damage to people on the ground."

"For older junk that we can't send instructions to, we could use special spacecraft to grab them in space and send them back to Earth. We might even be able to nudge them with lasers and send them back to Earth that way!"

Thanking Tai and their team, Mimi and the group left them to their important work. Mimi was very relieved that such a big problem had people like Tai working on it!

Over the next few days, Mimi and her new friends learnt everything that Mission Control could offer them! They met scientists, engineers and computer programmers; doctors, lawyers and even the chef who designs the astronaut's food! For every single one, Mimi had a whole bunch of questions and got more and more excited about launch day...

Chapter Nine

Reception

Chapter 9

It was launch day! Mimi could barely contain her excitement as they travelled from their hotel to Mission Control. Bouncing on her seat in the car, she craned her neck trying to see the spacecraft on the launchpad off in the distance. When they got to the front door of Mission Control, things were not quite as she expected...

Instead of the calm, friendly atmosphere she had got used to, it was in a state of panic! People were shouting and running around in all directions. Some of the more sensible ones were saving their breath and just stood still and shouted. Luna jogged up to meet them and she looked very concerned.

“Oh Mimi! It’s just awful!” said Luna.

“What’s happened?!” asked Mimi, imagining all sorts of catastrophes. Was something going to stop the launch?!

"It's the astronauts!" said Luna. "They're refusing to get onto the spacecraft!"

Mimi was confused. This was not what she thought Luna was going to say!

"They've all mislaid their personal items for the mission! Apparently, they think they've lost them somewhere in Mission Control and they've said that they can't possibly go into space without them!!!" Luna sighed.

"Well... What are they? Maybe we could help find them!" offered Mimi.

"That's the problem – we don't even know what they were planning to take with them! And they're all in a huff and are refusing to talk to us now! The only clues we have are in their mission profiles..."

Tanja

Name: Tanja

Role: Climate Scientist

Main Duties: A climate scientist looks at the atmospheres of other planets and does science in space to see what they might teach us about life here on Earth.

Fun Fact: Before becoming an astronaut, Tanja worked in Antarctica where she got to know her favourite animals – penguins!

Name: Anders

Role: Pilot

Main Duties: The mission pilot works closely with the Mission Control team to make sure that the spacecraft stays on course and keeps everyone safe whilst they are on their journey.

Fun Fact: Anders' favourite colour is bright pink and his hobby is entertaining his family and friends by singing and playing music.

Name: Amir

Role: Mission Commander

Main Duties: The mission commander makes sure that everyone is doing their job properly and is keeping safe. Amir is also the flight doctor and makes sure that everyone is healthy.

Fun Fact: Amir's family were worried that he might get cold in space so they have made him a cosy present to make sure he keeps warm and is reminded of them.

Name: Pixie

Role: Astrophysicist[5]

Main Duties: The mission astrophysicist studies the behaviour of planets, stars and other objects. Whilst in space, Pixie will be doing experiments to find out more about what different things in the solar system are made from.

Fun Fact: Pixie is very competitive and she loves playing sports. She's looking forward to starting some game tournaments with her crewmates!

---

[5] *Say "Astro-fizzy-sist" – a type of space scientist*

Mimi looked at the mission profiles and thought to herself carefully...

MIMI NEEDS YOUR HELP!

Can you search through the pictures of Mission Control and find the personal items that the astronauts have lost?

Look for the things that don't belong and see which one matches each of the astronauts! And be quick, the mission depends on you!!!

Mimi thought hard about her visit to Mission Control and she knew that some things didn't seem quite right! At top speed, she raced down the corridors of Mission Control with her trainers squeaking on the floor. She found the missing items, one after the other, in the places she had visited and, with her arms full of precious cargo, she returned panting to the control room.

CHAPTER TEN

Chapter 10

With her arms full of the astronauts' lost property, Mimi entered the control room and everyone in the room turned to her and went silent.

"Umm... Ta-da?" she said.

Immediately, all of the Mission Control team started cheering loud enough to make Mimi jump. Thanking her over and over again, an official took the astronauts' items and rushed with them out through a doorway that led to the launchpad. Mimi was then swamped by relieved-looking people who all wanted to thank her, shake her hand or pat her on the back – she had saved the day!

Luna raced over to her and gave her a very enthusiastic high-five.

"Mimi! You did it!!!" she cheered. They went to their VIP seats and had a celebratory glass of juice as they tried to calm down after all the excitement. All of the Mission Control team had immediately gone back to work and were getting ready for the launch, speaking into radios, sharing information and working at their computers.

The giant countdown clock on the wall was ticking down and it was almost time for launch!

Someone came up to Luna and whispered in her ear. Luna nodded excitedly and grabbed Mimi's hand.

"Mimi, come quickly! We've got one more job for you!"

"Wha-?" Mimi managed before they rushed away.

A seat had been cleared at the front of the room with a microphone and a big red button on a desk in front of it.

"Mimi, because you have saved the day, this honour is yours! Start the final countdown and, at zero, push the big red button for launch!" said Luna, proudly. Granny and 'Uncle' Colin looked on with amazement as Mimi took her seat...

In a big, brave voice, Mimi started the countdown...

"10... 9... 8... 7... 6... 5... 4... 3... 2... 1... **LIFT-OFF!!!**" and pushed the button!

On the big screen, she saw the rocket's engines blasting into life and the rocket started to rise up and up into the sky...

"Woof!" said 'Uncle' Colin, impressed.

"Bless you!" said Luna, offering him a tissue.

A FEW WEEKS LATER

The summer holiday was nearly over. Mimi was packing her bag to go home to her mum and dad and the start of school. She heard the letterbox rattle and Colin, woofing excitedly, ran to the door.

He came into her room with a parcel in his mouth.

"Mmmmph." Woofed Colin, his mouth full.

She unwrapped it carefully and read the note that was inside.

Dear Mimi,

Thank you so much again for saving our space mission!

The astronauts didn't get a chance to thank you in person but they send their best wishes and they are having a great time in space – all thanks to you!

We hope you like this little present from them.

Your friend,

Luna

She turned over the object in her hand to see a photo frame with a picture of the smiling astronauts, all floating in zero-gravity and holding the precious items that she had found for them! She hugged the photo and put it carefully in her suitcase on top of her neatly-folded astronaut's jumpsuit.

She was going to have a LOT to talk about with her friends at school in a few days! She was now **Mimi: Space Adventurer!**

The End

Meet Some Real-Life Science Super Stars!

Alex

Name: Alex Davies

Job: Systems Engineer at RAL Space

What Do You Do? I make sure satellites will keep working, even when they are far away in Space. In my lab, I use vibration to check it does not shake apart during launch, I check it can survive hot and cold temperatures, work without any air, and more! If the satellite survives my tests, then it is allowed to launch and go up into Space.

What three words would you use to describe yourself? Shy, ambitious, fun.

How did you get to where you are now? It's been a long journey! I moved jobs a lot before I found this job. I have worked in Consultancy, Printing, and Nuclear before I decided to apply for a Space job. And one day, I might move again. I have a degree in a STEM subject, so it is very easy to move between jobs!

What inspired you to choose your job? When I was a child, I got some Space books for Christmas, with pictures of galaxies swirling around each other, lit up like fireworks in the black of the Universe. For about 10 years I did other jobs, and when I wanted a new job, I remembered those images in my old books, and it inspired me to apply for what I do now.

What is the best thing about what you do? I love learning! I am thankful to be in a job where I am still learning! I have a mentor, who is an expert at what they do, who teaches me. And I mentor to someone else, too, so I can pass my knowledge along to them.

Name: Ana Paula Castro de Paula Nunes

Job: Heat Shield Product Lead at Space Forge

What Do You Do? I help develop the technology to return our satellites from space safely back to Earth.

What three words would you use to describe yourself? Ambitious, stubborn, playful.

How did you get to where you are now?
It was a long ride. I am originally from Brazil, but I always wanted to study and work abroad. First, I had to learn English at school! It took me three attempts to get accepted to study Aerospace Engineering, but midway through my degree I got a scholarship to study in the UK. I then got selected to go to China to study Space Law and Policy. It was an amazing and different experience, and the decision to go to China also brought me opportunities like working for the United Nations in Vienna and becoming an analog space astronaut in a mock-up lunar mission in a volcano in Hawaii! After all these life-changing experiences, I knew I wanted to continue working in the space sector. I found Space Forge and fell in love with what the company is trying to achieve!

What inspired you to choose your job? My excitement for space, adventures and wanting to work in a place that is helping Earth become a better place for future generations.

Is your job hard or dangerous or fun? It is both hard and fun. Once we send the satellites to space (the hard part), bringing them back to Earth will be very fun!

Name: Ben Dryer

Job: Space Instrumentation Research Fellow at The Open University

Ben

What Do You Do? The Sun emits dangerous particles, on Earth we are shielded from them by our magnetic field, but in space those particles can damage equipment. I help to understand how these particles damage cameras that take pictures of other planets, so we get the best images possible back from space.

What three words would you use to describe yourself? Curious, approachable, collaborative.

How did you get to where you are now? I didn't know what I wanted to when I finished school. I was passionate about teaching, and good at physics. I studied physics at university but wasn't sure if it would lead to a career. I spent a year in industry where I became interested in doing a physics related PhD, and that grew into a research fellowship where I get to work on space missions and teach space science. It's a perfect balance of all the things I was interested in when I was younger, although I didn't plan to end up here!

What is the best thing about what you do? Things I work on go into space! I work with large international teams of scientists, and we have worked on cameras that have flown on major ESA and NASA missions.

If you were going to space, what is the one personal/luxury item you would bring? My kindle with a selection of books (mostly sci-fi!)

Irene

Name: Irene Di Giulio

Job: University Lecturer at King's College London

What Do You Do? My research is quite varied, it spans from human space exploration to human movement in health and disease. I study advantages and needs of astronauts with physical disabilities to make sure that they can be safe and healthy in space. I am fascinated by how gravity shapes human development and how people can adapt to it.

Hobbies: Sports, reading, dancing, going to the theatre.

What three words would you use to describe yourself? Energetic, creative, passionate.

What inspired you to choose your job? I liked science and I wanted to learn about new ways to improve the world around me.

What is the best thing about what you do? The people I have met. I am lucky enough to have worked in different labs and I have met fantastic people from all over the world. My students are great and when travelling for conferences or other activities I met inspirational people.

What are the people you work with like? The people I work with are driven and extremely intelligent. There is quite a lot of collaboration, and the environment is really supportive.

If you were going to space, what is the one personal/luxury item you would bring? I know that it would not work, but a proper shower!

Isha

Name: Isha Parvaiz

Job: Physics Student at King's College London, previous Intern at The European Space Agency

What Do You Do? I've just finished doing an internship with the European Space Agency. I was working on making a new way for astronauts to learn how to use a robotic arm! I used virtual reality, which is where you put on a headset, see images, and interact with your surroundings. It was all very fun, and I worked with amazing people, including astronauts, robots, engineers and much more!

Hobbies: Playing piano, reading books, hiking.

What three words would you use to describe yourself? Outgoing, diligent, enthusiastic.

What is the best thing about what you do? During my internship, the best thing was meeting so many cool people all the time! I met astronauts and astronauts in training, along with people who have worked on brilliant projects and seen amazing things like rocket launches. I loved interacting with these people daily, getting to know more about them and hear their inspirational journeys.

What are the people who work with you like? The people in my internship team were all students like me, so we got on well! We liked hanging out together and helping each other - we worked well together. Our supervisor was great and extremely chill! He encouraged us to push the boundaries of our work at ESA, and he supported us in a great way.

Name: Jamie Lloyd

Job: Electrical & Avionics Group Manager at Astroscale

Jamie

What Do You Do? I help design and build the satellites that remove junk from space. I pick the right electrical building blocks such as computers, power supplies, and radios that are needed for the satellite to think, speak, and operate and how these parts can talk to each other for the satellite to work.

What three words would you use to describe yourself? Practical, thoughtful, laid-back.

How did you get to where you are now? I have always loved space, as a kid (and as an adult!). Before university I managed to gain a year work experience for a company making the power systems for jet planes. This led me to study Electrical Engineering with Space Science at university. Once I graduated, I joined a satellite company and worked on all sorts of satellites, from small earth observation ones up to very large GPS satellites. I got to work with NASA on a science mission as well which was hugely exciting. Finally, I had an opportunity to join Astroscale where our focus is on helping clean up the rubbish we have left in space.

What is the best thing about what you do? Looking after the space environment (as we are trying to do here on Earth) before it's too late. Space and satellites are so important to our everyday life, so if we need to keep space clean and clear for future generations.

If you were going to space, what is the one personal/luxury item you would bring? Either my phone, or a Star Wars movie (watching that in space would be amazing).

Name: Josh Gribben

Job: Analog Astronaut and PhD Researcher at the University of Strathclyde

Josh

What Do You Do? I look for ways to make space more useful for people on Earth and help future astronauts on the Moon and Mars talk to people back home.

Hobbies: Playing video games and hiking in nature.

What three words would you use to describe yourself? Enthusiastic, relaxed, determined.

How did you get to where you are now? I went back to education to follow my dream of working in space. First, I went to college to study physics and maths before continuing into an aircraft engineering course. I then studied Aerospace Systems at the University of Glasgow. Since no one in my family had been to university before everything about it was new! Finally, I started my PhD at the University of Strathclyde and took part in an Analog Astronaut mission.

What is the best thing about what you do? I love all the exciting opportunities to help humans explore and use space. When I did my Analog Astronaut mission, I got to work together with people from all over the world to explore the kinds of problems faced by real astronauts in space and how we can try to solve them, it was an incredible experience!

What are the people you work with like? Inspiring. Our team is trying to address global challenges using space technology, and they are all very passionate about their work.

Name: Jordan Grant

Job: Lead Manufacturing Engineer at Space Forge

What Do You Do? I lead a team of engineers who are responsible for building the satellites that we send into space!

Jordan

What three words would you use to describe yourself? Driven, approachable, reliable.

How did you get to where you are now? I started out as an apprentice for Aerospace Engine Manufacturer Rolls-Royce, where I learned a lot about the core engineering, maths, and physics. I really enjoying learning how things were made and getting to make things myself! I got to travel around the UK and Germany working in many different places with lots of amazing people. After 10 years I decided to try something different and thought about what I love... that was space! I started as a manufacturing engineer for Space Forge in 2022 and have had SO much fun since!

What is the best thing about what you do? I get to work with some great people, on things we are all so passionate about - and we are doing something that is just so cool!

Is your job hard or dangerous or fun? Yes! It is important to always think through what you are doing carefully to make sure you are safe, that is the most important thing. Once you know what you are doing is safe, it can be hard, but you can break it down into easier bite size chunks which also makes it incredibly fun!

Name: Kirsty Lindsay

Job: Assistant Professor of Rehabilitation Sciences at Northumbria University

Kirsty

What Do You Do? Living in space isn't very good for you: so, I research how to keep astronauts healthy. I look at the forces and accelerations that astronauts experience when they exercise, as it is different to when they are on Earth.

Hobbies: Gaming (especially Pokémon and Legend of Zelda games!), crochet, spinning, having picnics with my little girl.

How did you get where you are now? I started out fixing jet engines and did a mechanical engineering apprenticeship, then I hurt my leg so I needed a new job. I decided to be a physiotherapist, and after working for a little while in the NHS decided to take master's degree. I studied Space Physiology and Health. After I finished, I was really lucky to get a job at the European Space Agency. When I came back to the UK, I wasn't sure what to do again, but a PhD sounded fun, so I did that. In my last year I was asked to start a physiotherapy clinic at Northumbria University and do some research about exercising in space. Now I run a physio clinic and do space and health research, with one focus being on how to keep astronauts healthy on the Moon!

What is the best thing about what you do? I love doing experiments and finding out new things no-one else knows yet, and I get really excited when I get to talk to people about my work.

Name: Meg Schwamb

Job: Astronomer, planetary scientist, and reader at Queen's University Belfast

What Do You Do? II use telescopes to study how planets are formed and evolve. I study our Solar System and planets around other stars in our galaxy.

What three words would you use to describe yourself? Passionate, hardworking, funny.

How did you get to where you are now? I grew up in the United States of America (USA). I studied physics in university, then went on to get a PhD in planetary science and I have lived all across the USA and in Taiwan. Eventually, I made my way to Hawaii and worked for three years at the Gemini Observatory, which has a telescope on the dormant volcano called Maunakea! I moved to Northern Ireland in 2019. I now work at Queen's University Belfast teaching physics and astronomy to undergraduate students. I mentor a research group and do my own research too.

What inspired you to choose your job? Ever since I've been a little girl, I've been fascinated by the night sky and studying our Solar System. I learned what an astronomer was, and that it was something I could do, after reading a book by Stephen Hawking.

Is your job hard or dangerous or fun? Most days, I have fun learning new things, plotting up new data, and figuring out what my results mean. It's hard when things don't go to plan or there's an obstacle to overcome, but that's the fun and challenge of being a scientist.

Name: Melissa Brobby

Job: Science Communicator / Writer / Digital Communications Officer at the UK Space Agency

Melissa

What Do You Do? In my day job, I promote the work of the UK Space Agency and what is happening in the UK space sector on social media. Outside of my job, I write articles on space-related subjects from spaceflight to the Solar System and beyond for Sky at Night magazine, and I am a judge for the Royal Observatory's Astronomy Photographer of the Year award.

Hobbies: Stargazing, baking, dancing, travelling.

What three words would you use to describe yourself? Curious, enthusiastic, ambitious.

How did you get to where you are now? I worked as a journalist in my 20's before transitioning to writing content online. When Maggie Aderin-Pocock was announced as a presenter for Sky at Night, it showed me that someone who looked like me could do the very thing I had always wanted to do – share the wonders of space with a wider audience. I decided to embark on a career change, and in my mid 30s I studied Astronomy, and I now create content for the UK Space Agency's social media and write articles for Sky at Night magazine.

What inspired you to choose your job? I have always loved writing and answering people's space questions, so my job is the perfect combination of my two passions.

If you were going to space, what is the one personal/luxury item you would bring? My iPod and headphones.

Name: Patrick Harkness

Job: Professor of Exploration Technology at Glasgow University

Patrick

What Do You Do? I work to build better scoops and tools for space rovers. The soil on other planets is not quite the same as we're used to on Earth, so we need special tools to take samples. These samples then help us to better understand what the other planets are like, and if we could live there some day.

What three words would you use to describe yourself? Cheerful, supportive, determined.

How did you get to where you are now? I think it is important to have a long-term objective and try to move a little closer to it every day. Some days you make more progress than others.

What inspired you to choose your job? I always wanted to be involved in space exploration, and I think I chose a career in academia because it provides a lot of opportunities to make that happen.

What is the best thing about what you do? I get to go to cool places to test our space exploration equipment, such as Antarctica, which is where my picture is taken. Parts of Antarctica are quite Mars-like, and if our hardware will work there, it has a good chance of working in space as well.

If you were going to space, what is the one personal/luxury item you would bring? Mint ice cream.

Peter

Name: Peter Shaw

Job: Rocket Scientist a.k.a Senior Lecturer in Space Technology at Kingston University

What Do You Do? I am an award winning proud dyslexic rocket scientist, who teaches young adults how to design, build and test fire: rocket engines used to get vehicles into space, and space propulsion systems that spacecraft use to get around whilst in space. My vision is to help grow the UK space launch sector.

Hobbies: Climbing up cliffs, listening to rock music, playing with my kids, and dressing up like a goblin and pretending to be in a medieval fantasy world.

What three words would you use to describe yourself? Inquisitive, cheeky, daring.

What inspired you to choose your job? A mixture of watching Sci-fi (Star Trek, Star Wars etc) helped to inspire my imagination and a desire to help the next generation inspired me to become a university lecturer. I've always been good at explaining complex subjects and I think teaching others is my life calling.

What is the best thing about what you do? I get to build rocket engines and test fire them in my laboratory, learning how to make them work better. Ultimately, I want to build a space propulsion system to send humans to Jupiter, but I'm a long way off that yet.

Is your job hard or dangerous or fun? Yes to all three. It's rocket science = hard! Things can go boom = dangerous! It's fun because things go boom!